Unlocking the Power of the Sun: Agrivoltaics and Sustainable Agriculture

By Ben Davis

Contents

Chapter 1: Introduction to Agrivoltaics

As the world grapples with the urgent need to transition towards sustainable agricultural practices, researchers and farmers have been exploring innovative ways to maximize land use and minimize environmental impact. One such practice that has gained increasing attention is agrivoltaics, a groundbreaking approach that combines solar energy production with agricultural activities. This chapter provides an overview of agrivoltaics as a sustainable agricultural practice and explores its potential benefits and challenges.

Agrivoltaics, also known as agrophotovoltaics, involves the co-location of solar panels and agricultural crops on the same land. By harnessing the power of the sun for both electricity generation and crop cultivation, agrivoltaics offers a promising solution to address the pressing issues of land scarcity and climate change. In traditional solar farms, vast expanses of land are dedicated solely to energy production, depriving agricultural activities of much-needed space. Agrivoltaics, on the other hand, allows a harmonious integration of agriculture and energy generation, making efficient use of land resources.

The benefits of agrivoltaics are manifold. Firstly, it provides a dual income stream for farmers. By leasing their land for solar panel installation, farmers can not only continue their traditional agricultural practices but also benefit financially from the electricity generated. This additional income can help enhance the resilience of farming communities and support their transition to more sustainable practices. Moreover, agrivoltaics reduces the risk of relying solely on crop production, as the solar panels guarantee a stable source of revenue.

In addition to economic advantages, agrivoltaics offers significant environmental benefits. The strategic placement of solar panels in agrivoltaic systems provides shade for crops, reducing evaporation and water requirements. The shade also creates a more favorable microclimate, mitigating the adverse effects of extreme temperatures on plants. This translates into increased crop yields and improved resilience in the face of climate variability and change. Furthermore, agrivoltaics promotes biodiversity by creating new habitats for pollinators and other beneficial organisms, fostering ecological balance within agricultural landscapes.

Despite the numerous benefits, agrivoltaics also presents challenges that need to be addressed. The design and management of these integrated systems require careful planning to avoid shading crops excessively or hindering energy production. Furthermore, optimizing the spacing between solar panels and crops is crucial to ensure efficient light distribution for both. The selection of suitable crops for agrivoltaic systems is another consideration, as not all crops thrive under shade conditions. These challenges necessitate ongoing research and collaborative efforts between experts in agriculture, energy, and environmental sciences.

In conclusion, agrivoltaics represents a compelling solution for sustainable agriculture and renewable energy production. By harnessing the power of the sun through the co-location of solar panels and crops, agrivoltaics maximizes land use, provides additional income for farmers, and offers numerous environmental benefits. However, to fully unlock the potential of agrivoltaics, further research and development are needed. The second half of this chapter will delve deeper into the technical aspects of agrivoltaic systems and explore successful case studies that highlight their tremendous potential. Stay tuned for the next part of this enlightening journey.Agrivoltaics is a transformative practice that holds great promise for the future of sustainable agriculture and renewable energy production. In the first half of this chapter, we explored the concept of agrivoltaics, its benefits, and the challenges it presents. Now, let us delve deeper into the technical aspects of agrivoltaic systems and uncover successful case studies that highlight its tremendous potential.

One of the key considerations in implementing agrivoltaics is the design and management of these integrated systems. Careful planning is essential to ensure that the solar panels do not excessively shade the crops or hinder energy production. Optimal spacing between the solar panels and crops plays a crucial role in maintaining efficient light distribution for both elements of the system. This necessitates continuous research and collaboration between experts in agriculture, energy, and environmental sciences.

Effective management of agrivoltaic systems also involves selecting suitable crops for cultivation under shade conditions. While some crops thrive in the presence of shade, others may struggle to meet their full growth potential. Researchers and farmers are working together to identify and develop crop varieties that are well-suited for agrivoltaic systems. Through ongoing experimentation and innovation, they aim to maximize the productivity and resilience of these integrated systems.

To further explore the potential of agrivoltaics, let us take a closer look at successful case studies. One inspiring example comes from the Kyushu Institute of Technology in Japan. They have been experimenting with agrivoltaic systems that combine solar panels, vegetable production, and aquaculture. By carefully arranging the solar panels to create optimal microclimates, they have achieved remarkable results. The shade provided by the solar panels has improved crop yields by reducing water requirements and preventing overheating during hot summer days. Moreover, the system has created a favorable environment for the cultivation of fish and other aquatic organisms, thereby enhancing overall productivity.

Another noteworthy case study takes us to the mountains of Switzerland, where the Fraunhofer Institute for Solar Energy Systems has implemented agrivoltaic systems in alpine pastures. By installing solar panels above grazing areas, they have not only generated renewable energy but also improved forage quality for livestock. The solar panels provide shade to the pasture and contribute to the creation of a more favorable microclimate, resulting in increased grass growth and nutritional value. This innovative approach ensures the sustainable use of alpine land while contributing to the local energy transition.

These success stories demonstrate the immense potential of agrivoltaics to revolutionize agricultural landscapes around the world. By harnessing the sun's energy through integrated systems, we can address the challenges of land scarcity, climate change, and food security simultaneously. Agrivoltaics offers a pathway to a more sustainable and resilient future for both farmers and the global community.

As we conclude this chapter, we have explored the concept, benefits, challenges, and successful case studies of agrivoltaics. The integration of solar panels and crops in a harmonious and efficient manner maximizes land use, provides financial opportunities for farmers, and offers numerous environmental benefits. However, it is crucial to acknowledge that agrivoltaics is still an evolving field, and there is much more to be explored and discovered.

In the chapters that follow, we will continue our journey through the world of agrivoltaics. We will delve into the technicalities of system design, explore potential policy frameworks, and examine the economic viability of agrivoltaic systems. Join us as we uncover the intricacies and potentials of agrivoltaics, and let us together unlock the power of the sun for a sustainable and prosperous future.

Chapter 2: Understanding Solar Energy

The sun, that radiant ball of fiery energy that illuminates our world, has been a source of fascination for centuries. Its power is immense, and we are only just beginning to fully comprehend its potential. In this chapter, we will embark on a journey to understand solar energy and its crucial role in agrivoltaics - the marriage of solar power and sustainable agriculture.

Solar energy is harnessed by utilizing photovoltaic (PV) cells, which are commonly found in solar panels. These cells are made up of semiconductor materials, such as silicon, that have the remarkable ability to convert sunlight directly into electricity. When sunlight hits the surface of a PV cell, it excites the electrons in the material, causing them to flow and create an electric current. This clean and renewable form of energy generation has revolutionized the way we approach power production.

The importance of solar energy in agrivoltaics cannot be overstated. By installing solar panels on agricultural land, we not only generate clean electricity but also provide shade to crops, creating a mutually beneficial relationship. This practice, known as agrivoltaics or dual-use farming, optimizes land use and energy production, while simultaneously enhancing agricultural productivity.

In agrivoltaics, the role of solar energy extends beyond electricity generation. The shaded environment created by the solar panels reduces the intensity of sunlight reaching the plants, shielding them from excessive heat and minimizing evaporation. This can be particularly advantageous in arid regions, where water scarcity poses a significant challenge to farming. By mitigating the effects of intense sunlight, agrivoltaics helps conserve water and improve crop yields, promoting sustainable agriculture.

Additionally, solar panels offer another avenue for resource efficiency in agrivoltaics. As solar panels require a stable operating temperature to function optimally, they actively cool down as they convert sunlight into electricity. This cooling effect is advantageous for both the panels and the crops beneath them. By reducing the temperature of surrounding air, evaporation rates decrease, providing further water conservation benefits for crops in need.

Furthermore, solar panels often raise the height of crops due to their mounting structures. This additional vertical space can be utilized to grow shade-tolerant crops or support vertical farming systems. By making use of otherwise underutilized airspace, agrivoltaics maximizes land utilization and opens up possibilities for diversification of crop production.

As we delve deeper into the fascinating world of agrivoltaics, we will explore the intricate mechanisms behind this sustainable farming practice. In the second half of this chapter, we will investigate its environmental benefits, economic feasibility, and the various innovative techniques employed in the implementation of agrivoltaic systems.

But for now, let us bask in the marvel that is solar energy. Its ability to power our world while nourishing our crops is nothing short of extraordinary. As the sun continues to rise and set, it bestows upon us an abundance of energy waiting to be unlocked and harnessed. Join us on this captivating journey to unlock the power of the sun and unlock a brighter, more sustainable future.In the second half of this chapter, we will delve into the environmental benefits, economic feasibility, and innovative techniques that make agrivoltaics a compelling and sustainable farming practice.

One of the key environmental benefits of agrivoltaics lies in its ability to reduce water consumption. As mentioned earlier, the shaded environment created by solar panels helps minimize evaporation, thus conserving water. This is particularly crucial in arid regions where water scarcity poses a significant challenge to agriculture. By mitigating the effects of intense sunlight, agrivoltaics provides a shield to crops, reducing water loss and improving overall crop yields. This water-saving aspect of agrivoltaics not only promotes sustainable farming but also addresses a pressing global concern.

Furthermore, incorporating solar panels into agricultural land offers significant economic benefits. Alongside crop production, farmers can generate clean electricity for their own use or sell it back to the grid. This provides an additional revenue source and reduces reliance on fossil fuels, leading to long-term cost savings and increased energy independence. Agrivoltaics also opens up opportunities for collaboration between farmers and solar energy companies, allowing for innovative partnerships and mutually beneficial agreements.

In terms of implementation, agrivoltaics can take various forms. In some setups, solar panels are mounted high above the crops, allowing for sunlight to permeate both horizontally and vertically. This arrangement provides a favorable microclimate for crops, creating a partially shaded environment and maintaining suitable light conditions for plant growth. Other systems may have solar panels mounted at a lower height, providing greater shade and protection against intense sunlight. The choice of system depends on the specific crop requirements, climate conditions, and regional factors. Farmers and researchers are continuously exploring and optimizing different designs to maximize crop productivity and energy generation.

Additionally, advancements in technology and innovation have paved the way for exciting developments in the field of agrivoltaics. For example, researchers are exploring the use of transparent solar panels, which allow a certain portion of sunlight to pass through while simultaneously generating electricity. This innovative approach maintains adequate light levels for plant growth while harnessing solar energy, offering a potential solution for crops that require higher levels of direct sunlight.

Agrivoltaics also opens up avenues for integrating other sustainable farming practices. For instance, the shaded environment created by solar panels can facilitate the cultivation of shade-tolerant plants or the implementation of vertical farming systems. This synergy allows farmers to diversify their crop selection and optimize land utilization, fostering crop rotation and promoting biodiversity.

In conclusion, agrivoltaics harnesses the power of solar energy to revolutionize agriculture, benefiting both the environment and the farmers. By providing shade to crops, reducing water loss, and generating clean electricity, this innovative practice ensures a sustainable and resilient future for food production. With ongoing research, technological advancements, and the growing global commitment towards sustainability, agrivoltaics is poised to play a significant role in shaping the agricultural landscape. So let us seize this opportunity, unlock the power of the sun, and pave the way for a brighter and more sustainable future in agriculture.

Agrivoltaics, an innovative approach that combines solar panels and agricultural practices, holds immense promise for sustainable agriculture. The integration of solar energy infrastructure amidst growing crops not only maximizes land use efficiency but also offers a range of additional benefits. In this chapter, we delve deeper into what agrivoltaics entails and explore its potential to revolutionize the way we harness energy and cultivate food.

At its core, agrivoltaics aims to optimize the coexistence of solar photovoltaic (PV) systems and agriculture, allowing solar panels to be installed above the agricultural fields. By providing shade to the crops, solar panels help reduce evaporation, maintain soil moisture levels, and mitigate extreme temperature fluctuations. This synergy creates a microclimate that promotes plant growth and can extend the growing season, leading to increased crop yields.

In addition to its impact on agricultural productivity, agrivoltaics addresses many challenges associated with traditional solar energy generation. The practice of utilizing agricultural land for solar installations eliminates the need for extensive land acquisition, minimizing the potential disruption of natural habitats. By repurposing already cultivated land, agrivoltaics ensures that food production does not compete with renewable energy generation for land resources.

Moreover, combining solar power with farming activities offers significant economic advantages. Farmers can generate additional revenue by leasing their land to solar energy developers and becoming prosumers of both food and energy. With the right infrastructure and support, agrivoltaics can contribute to rural economic development, offering a sustainable income stream for farmers while reducing their reliance on unpredictable agricultural prices.

Furthermore, agrivoltaic systems have the potential to enhance the resilience of agricultural ecosystems. The shade provided by solar panels helps protect crops from harsh weather conditions such as hailstorms or excessive sunlight. This protective shield not only safeguards the plants but also reduces the risk of soil erosion, particularly in areas prone to heavy rainfall or high winds. By mitigating these risks, agrivoltaics can contribute to more stable and secure food production systems.

Beyond the immediate benefits, agrivoltaics also contributes to mitigating climate change. Solar panels generate clean, renewable energy, reducing greenhouse gas emissions associated with fossil fuel power plants. The integration of agrivoltaics on a larger scale has the potential to make a substantial impact on carbon emissions, helping to combat global warming and create a more sustainable future for our planet.

As we have explored in this first half of the chapter, agrivoltaics offers a compelling solution to combine the advantages of renewable energy and sustainable agriculture. With its potential to enhance crop productivity, foster economic growth, protect ecosystems, and combat climate change, agrivoltaics has gained increasing attention from researchers, policymakers, and farmers alike.

Come, let us venture further into the world of agrivoltaics. In the second half of this chapter, we will examine successful case studies, explore the technical aspects of implementing agrivoltaic systems, and discuss the challenges and future prospects of this innovative practice. Continue reading, and you will unlock the full potential of agrivoltaics and discover how harnessing the power of the sun can revolutionize sustainable agriculture.In the second half of this chapter, we will delve deeper into the world of agrivoltaics, exploring successful case studies, technical aspects of implementation, as well as the challenges and future prospects of this innovative practice. Through these insights, we will further unlock the full potential of agrivoltaics and discover how harnessing the power of the sun can revolutionize sustainable agriculture.

To truly grasp the efficacy of agrivoltaics, it is imperative to examine successful case studies from around the world. One such pioneering initiative can be found in Japan, where farmers have been integrating solar panels into their agricultural practices with remarkable results. This synergistic approach has not only allowed for land efficiency but has also significantly increased agricultural productivity. In this case, solar panels were installed atop mushroom farms, providing the dual benefit of shade for the delicate mushrooms and generating renewable energy simultaneously. This innovative practice not only optimizes land utilization but also offers a sustainable income stream for farmers, promoting rural economic development.

Another noteworthy case study can be witnessed in India, where agrivoltaics has been employed as a solution to address water scarcity in arid regions. By installing solar panels above the fields, farmers have successfully reduced crop water requirements through the provision of shade. The panels prevent excessive evaporation, preserving soil moisture and minimizing water usage. This approach not only fosters sustainable agriculture but also enhances water conservation efforts, thus creating a more resilient and resource-efficient farming system.

While case studies showcase the successful implementation of agrivoltaics, understanding the technical aspects of this practice is equally crucial. The integration of solar panels with agriculture requires careful planning and consideration. Factors such as crop selection, spacing, panel orientation, and height are essential to ensure optimal energy generation and crop growth. Appropriate monitoring systems must be implemented to evaluate the effectiveness of agrivoltaic systems and make necessary adjustments accordingly. Additionally, ongoing research is devoted to optimizing the arrangement of solar panels to maximize energy yield while maintaining agricultural productivity.

In spite of its immense potential, agrivoltaics also faces several challenges that require addressing. One such hurdle is the initial investment cost associated with setting up the necessary infrastructure. While advancements in technology have led to a declining trend in the cost of solar panels, further support and financial incentives are required to make agrivoltaic systems more accessible to farmers. Additionally, there is a need for tailored training and education programs to equip farmers with the knowledge and skills needed to effectively manage agrivoltaic systems.

Looking ahead, the future prospects of agrivoltaics are promising. As interest in renewable energy and sustainable agriculture continues to grow, we can anticipate further research and development in this field. Enhancing the integration of agrivoltaics within existing agricultural practices holds the potential to optimize resource utilization, support climate resilience, and contribute to greenhouse gas emissions reduction. Collaborative efforts among researchers, policymakers, and farmers are essential in facilitating the widespread adoption of this innovative practice and forging a path towards a more sustainable future.

In conclusion, agrivoltaics represents an extraordinary opportunity to harmonize renewable energy generation and sustainable agriculture. Through the integration of solar panels within agricultural fields, we unlock countless benefits such as increased crop yields, economic growth for farmers, ecosystem protection, and mitigation of climate change. As we have explored in this chapter, agrivoltaics has garnered increasing attention due to its potential for transformation. By embracing this revolutionary approach, we can truly unlock the power of the sun and pave the way for a more sustainable and resilient agricultural future.

Unlocking the Power of the Sun: Agrivoltaics and Sustainable Agriculture

In the quest for a cleaner and more sustainable future, agrivoltaics has emerged as a promising solution. This innovative approach combines the power of solar energy with agricultural practices to create a harmonious synergy that benefits both energy production and crop cultivation. By utilizing the vast potential of the sun, agrivoltaics presents a paradigm shift in sustainable agriculture. This chapter aims to highlight the advantages of agrivoltaics, exploring its positive impacts on energy production, crop yield, and water conservation.

One of the key benefits of agrivoltaics is its ability to maximize land utilization. Traditionally, large tracts of land were dedicated solely to either solar energy generation or agriculture. However, agrivoltaics challenges this conventional division by integrating photovoltaic panels into agricultural spaces. By doing so, agrivoltaics transforms areas previously underutilized into highly productive landscapes. This approach allows farmers and solar power producers to coexist, making efficient use of available resources and promoting land conservation.

Energy production is greatly enhanced through agrivoltaics. Not only do farmers have the capability to produce crops, but they can also become clean energy producers. Solar panels installed above crops capture sunlight, simultaneously generating electricity while sheltering plants. This dual-purpose land use not only reduces the need for additional land but harnesses the untapped potential of agricultural areas for renewable energy production. As a result, farmers can become active participants in the clean energy revolution, contributing to a more sustainable future.

Moreover, agrivoltaics leads to increased crop yield and quality. The strategic placement of solar panels above crops provides shade and protection from harsh weather conditions such as excessive heat and heavy rainfalls. These panels act as a shield, minimizing crop damage and optimizing growth conditions. Additionally, the controlled environment created by agrivoltaic systems reduces evapotranspiration, preventing excessive water loss from plants. This leads to improved water-use efficiency, ensuring that crops receive an adequate water supply even during times of drought. Consequently, agrivoltaics not only enhances the quantity but also the quality of agricultural produce.

Water conservation is a pressing concern in agriculture, especially in regions prone to drought. Agrivoltaics offers a sustainable solution to this challenge. By incorporating solar panels into the agriculture landscape, agrivoltaics reduces water requirements through a phenomenon called microclimate regulation. The shade provided by solar panels helps to maintain a more stable microclimate, minimizing water evaporation from the soil. Furthermore, the elevated position of the panels allows for rainwater to flow more efficiently, enabling effective irrigation and preventing water runoff. This synergy between solar energy and agriculture ensures optimal water management, making agrivoltaics an eco-friendly choice for sustainable farming.

In conclusion, agrivoltaics represents a revolutionary approach to sustainable agriculture. Through the integration of solar panels into agricultural spaces, this innovative practice maximizes land utilization, enhances energy production, increases crop yield and quality, and promotes water conservation. The benefits of agrivoltaics are immense, not only for farmers but for the entire community. As we delve deeper into the second half of this chapter, we will explore additional advantages and delve into the mechanisms that make agrivoltaics a game-changer in the field of sustainable development. Stay tuned for the exciting continuation of our exploration into the power of agrivoltaics!One of the significant advantages of agrivoltaics is its positive impact on soil health and fertility. The integration of solar panels into agricultural spaces not only enhances energy production but also promotes sustainable soil management.

By shading the soil, the solar panels protect it from excessive sunlight, preventing soil temperature fluctuations and reducing water loss through evaporation. This regulation of soil moisture creates a more stable environment for microbial activity and nutrient absorption by plants. The shade provided by the panels also helps suppress weed growth, reducing competition for nutrients and optimizing their availability for crops.

Moreover, the presence of solar panels can lead to improved soil structure. Solar panels act as a canopy, reducing the impact of heavy rainfall on the soil surface. This prevents soil erosion and the loss of valuable topsoil, which is rich in organic matter and essential nutrients. Consequently, agrivoltaics helps maintain soil structure, fertility, and overall soil health.

Another aspect that sets agrivoltaics apart is its contribution to biodiversity conservation. Traditional solar farms are often seen as ecological deserts, lacking in vegetation and biodiversity. However, with the integration of photovoltaic panels into agricultural systems, there is an opportunity to create diverse and thriving ecosystems.

The spaces between and around the solar panels can be utilized for planting native wildflowers, grasses, or other beneficial plants. These plantings attract pollinators and provide habitat for beneficial insects, birds, and other wildlife. The increased biodiversity on agrivoltaic sites can enhance natural pest control, reducing the need for chemical pesticides and promoting a healthier ecosystem.

Furthermore, agrivoltaics presents a unique opportunity for community engagement and empowerment. With solar panels integrated into agricultural landscapes, farmers can become active participants in the renewable energy sector. By generating clean electricity through the solar panels above their crops, farmers have the potential to contribute to local energy grids, reduce reliance on fossil fuels, and create new avenues of income.

Agrivoltaic projects also have the potential to benefit local communities by providing employment opportunities and stimulating rural economies. The installation, operation, and maintenance of agrivoltaic systems require skilled labor, creating jobs in green industries. Additionally, the surplus energy generated through agrivoltaics can be shared with neighboring communities, providing access to clean and affordable electricity.

In conclusion, agrivoltaics offers numerous benefits that go beyond energy production and crop cultivation. By maximizing land utilization, enhancing energy production, increasing crop yield and quality, promoting water conservation, improving soil health, conserving biodiversity, and empowering communities, agrivoltaics presents a holistic and sustainable approach to agriculture.

As we conclude this chapter on the benefits of agrivoltaics, it is clear that this innovative practice has the potential to revolutionize our agricultural systems and contribute to a cleaner and more sustainable future. By harnessing the power of the sun and integrating it seamlessly into agricultural spaces, agrivoltaics offers a path towards resilient, efficient, and environmentally friendly food and energy production.

In the next chapter, we will delve deeper into the practical applications and implementation strategies of agrivoltaics. We will explore case studies, technological advancements, and policy considerations that are driving the adoption of agrivoltaics worldwide. Stay tuned as we continue our journey into unlocking the power of the sun through agrivoltaics.

Chapter 5: Challenges of Implementing Agrivoltaics

Examining the potential obstacles, this chapter addresses factors such as cost, land availability, and maintenance in implementing agrivoltaics.

As the world continues to seek sustainable solutions to meet the growing demand for food and energy, agrivoltaics has emerged as a promising approach. By combining agriculture with solar photovoltaic (PV) systems, agrivoltaics offers a unique opportunity to maximize land use efficiency and enhance resource sustainability. However, like any innovative concept, implementing agrivoltaics comes with its fair share of challenges.

One of the primary concerns associated with agrivoltaics is the cost involved in setting up the systems. Solar PV panels, despite the significant advancements in technology and reductions in prices over the years, can still be expensive. When it comes to agrivoltaics, the cost exponentially increases as there is a need for additional infrastructure to support the integration of farming and solar panels. Farmers and landowners may hesitate to invest in agrivoltaics due to the initial financial burden and uncertainties about long-term profitability.

Another challenge lies in land availability. Agricultural land is limited, and farmers rely on every inch to grow crops and sustain their livelihoods. Introducing solar panels on productive farmland raises concerns about potential reductions in available land for cultivation. This issue becomes even more critical in regions where land scarcity is already a significant problem. Careful planning and strategic placement of agrivoltaic systems are crucial to ensure minimal disruption to existing agricultural practices.

Maintenance is yet another obstacle that needs to be considered. While solar panels require regular cleaning and maintenance to operate optimally, adding agricultural activities to the mix complicates the maintenance process. Additional attention and care must be provided to both the crops and the solar panels. Farmers will need to invest time, resources, and expertise to manage the dual requirements effectively. Lack of proper maintenance can negatively affect the efficiency of both agricultural production and energy generation.

Moreover, navigating the regulatory landscape surrounding agrivoltaics poses further challenges. Existing policies and regulations might not adequately address this novel concept, making it difficult for farmers and investors to obtain the necessary permits and incentives. Policymakers need to work closely with experts and stakeholders to develop frameworks that encourage and support the adoption of agrivoltaics. Clear guidelines and streamlined procedures can help overcome these regulatory hurdles.

While these challenges may seem daunting, the potential benefits of agrivoltaics cannot be overlooked. By combining agriculture and solar energy, agrivoltaics has the power to revolutionize sustainable farming practices, addressing food and energy security simultaneously. However, to unleash its true potential, these obstacles must be understood and tackled effectively.

In the second half of this chapter, we will delve deeper into each of these challenges, exploring potential solutions and success stories from around the world. Stay tuned as we take you on a journey to unlock the full potential of agrivoltaics and its role in sustainable agriculture.

(Note: The second half of this chapter will be written in the next installment)Maintenance is a crucial aspect of implementing agrivoltaics, as it requires careful attention to both agricultural practices and solar panel upkeep. Balancing these two requirements can be challenging, but with proper planning and management, it can be achieved.

In agrivoltaics, the maintenance process becomes more intricate due to the combination of farming and solar energy generation. Solar panels need regular cleaning to maximize their efficiency, but this can pose difficulties when agricultural activities are integrated. Dust accumulation, crop debris, and pollen can reduce the panels' performance and hinder solar energy production. Additionally, the presence of crops around the panels can obstruct easy access for maintenance tasks.

To effectively address these challenges, farmers need to adopt innovative techniques and technologies. For example, automated cleaning systems can be used to remove dust and debris from solar panels, eliminating the need for manual intervention. These systems can be programmed to clean the panels at optimal times to minimize disruption to the crops. Farmers can also explore vertical or elevated panel installations, allowing for better access during maintenance while maximizing land use efficiency.

Alongside maintenance, navigating the regulatory landscape surrounding agrivoltaics can be a hurdle in its implementation. Existing policies and regulations may not adequately address this novel concept, creating barriers in obtaining necessary permits and incentives. To overcome this challenge, close collaboration between policymakers, experts, and stakeholders is vital.

Policymakers must work towards developing clear guidelines and streamlined procedures that facilitate the adoption of agrivoltaics. This involves creating frameworks that incentivize farmers and landowners to invest in these systems while ensuring they adhere to regulatory standards. Policymakers can also consider providing financial incentives, tax breaks, or grants to encourage the widespread implementation of agrivoltaics.

Success stories and best practices from around the world serve as valuable inspiration for overcoming these challenges. For instance, in Japan, where limited land availability is a concern, agrivoltaics have shown great potential. By strategically placing solar panels above crops that require shade, farmers have successfully increased their land use efficiency without compromising on agricultural productivity. This innovative approach demonstrates the possibilities of harmonizing solar energy production and agriculture.

In India, where water scarcity is a significant challenge, agrivoltaics have proven to be a beneficial solution. Farmers have integrated solar panels with drip irrigation systems, optimizing water usage while generating renewable energy. This dual-purpose system not only supports sustainable agricultural practices but also contributes to the country's renewable energy goals.

These success stories highlight the importance of knowledge sharing and collaboration between farmers, researchers, and policymakers. By learning from each other's experiences, stakeholders can collectively address challenges, develop efficient solutions, and create a supportive ecosystem for agrivoltaics.

In conclusion, agrivoltaics presents immense potential for sustainable agriculture and renewable energy generation. Despite the challenges of cost, land availability, maintenance, and navigating regulations, these hurdles can be overcome. By adopting innovative techniques, incorporating best practices, and working closely with policymakers, farmers and landowners can unlock the full potential of agrivoltaics. In doing so, we can create a future where agriculture and solar energy coexist harmoniously, addressing food and energy security while paving the way towards a more sustainable planet.

Chapter 6: Case Studies in Agrivoltaics

Over the years, agrivoltaics has proved to be a promising approach in harnessing solar energy while simultaneously promoting sustainable agriculture. This chapter presents real-world examples of successful agrivoltaic systems, showcasing their effectiveness in different regions and agricultural contexts.

One notable case study is the Graber Family Farm located in Oregon, USA. The Graber family had been struggling with the unpredictable weather conditions and decreasing crop yield in recent years. Seeking a solution, they decided to implement an agrivoltaic system in their fields. Rows of solar panels were installed above the crops, providing shade and reducing evaporation, ultimately resulting in more consistent moisture levels in the soil. By optimizing the use of available land, the Graber family not only increased their crop production but also significantly reduced water usage.

Moving across the globe to Portugal, the Teixeira Vineyard is another inspiring example. With a history of grape cultivation spanning generations, the Teixeira family faced a common challenge in the wine industry - excessive heat during the summer months. They integrated solar panels into their vineyard, strategically positioning them to provide shade for the delicate grapevines. This combination of solar energy and viticulture has not only protected the vines from the scorching sun but also increased the overall energy efficiency of the vineyard. The Teixeira Vineyard now proudly produces an excellent quality of wine while reducing its reliance on traditional energy sources.

In a completely different setting, the Kibbutz Ketura in Israel established an agrivoltaic system in their desert farmland. Facing limited water resources and intense sunlight, the kibbutz combined solar panels with date palm cultivation. The panels not only provided shade but also reduced water evaporation from the date palms, enabling the trees to thrive in such harsh conditions. This integration of solar energy and agriculture has not only increased food production in the region but also paved the way for sustainable farming practices in arid environments.

As we observe these diverse case studies, it becomes evident that agrivoltaics is not limited to a specific crop or geographical location. Its potential in revolutionizing sustainable agriculture is vast and adaptable to different farming practices. Whether it's fruits, vegetables, or vineyards, the integration of solar panels into agricultural landscapes offers numerous benefits.

Innovative projects continue to emerge worldwide, and the positive impact of agrivoltaics extends far beyond improving crop yield. It contributes to the reduction of greenhouse gas emissions, enhances water management, and promotes energy self-sufficiency in agricultural communities. These case studies exemplify the potential for agrivoltaics to revolutionize the way we approach food production and energy generation.

With that, let us delve deeper into the intricate mechanisms behind successful agrivoltaic systems. In the second half of this chapter, we will explore the technical aspects, implementation challenges, and future prospects for agriculovotaics. Stay tuned as we unravel the untapped potential of agrivoltaics and its transformative power in sustainable agriculture.In the second half of this chapter, we will delve deeper into the intricate mechanisms behind successful agrivoltaic systems. By exploring the technical aspects, implementation challenges, and future prospects of agrivoltaics, we can gain a comprehensive understanding of its transformative power in sustainable agriculture.

One crucial aspect of agrivoltaics is the careful design and placement of solar panels. The orientation and tilt of the panels must be strategically determined to maximize solar energy absorption while minimizing shading on the crops. Additionally, the height and spacing between the panels and the vegetation should be optimized to allow sufficient light penetration for plant growth. Balancing these factors requires a meticulous approach, considering various aspects such as crop type, geographical location, and available land.

Furthermore, the choice of crops plays a significant role in the effectiveness of agrivoltaic systems. Diverse cropping systems can be integrated, including vegetables, grains, and even livestock grazing. Selecting complementary crops that thrive in the shade provided by the solar panels enhances the overall productivity of the system. This approach not only ensures a diverse and sustainable food production but also maximizes the land-use efficiency.

Implementation challenges in agrivoltaics include technical considerations and financial viability. Developing a successful agrivoltaic system requires collaboration between agriculturists, solar energy experts, and researchers. Challenges such as engineering suitable mounting structures and optimizing energy production can be overcome through innovation and research. Additionally, financial incentives and government support play a significant role in encouraging farmers to adopt these sustainable practices.

Looking ahead, the future prospects for agrivoltaics are promising. With advancements in solar panel technology, we can anticipate more efficient and economical systems, further reducing greenhouse gas emissions and reliance on traditional energy sources. Additionally, ongoing research aims to optimize plant growth under partial shade conditions, maximizing crop yield and quality in agrivoltaic systems.

One particularly exciting development is the integration of energy storage technologies with agrivoltaic systems. By harnessing excess solar energy during the day and utilizing it during lower light periods or at night, farmers can enhance energy management and reduce reliance on the grid. Storage technologies, such as batteries, hold great potential for making agrivoltaic systems even more self-sufficient and economically viable.

In conclusion, agrivoltaic systems have proven to be a successful and adaptable approach in harnessing solar energy while promoting sustainable agriculture. The diverse case studies presented in this chapter highlight the vast potential of agrivoltaics, transcending geographical barriers and crop types. As we continue to explore the intricate mechanisms, implementation challenges, and future prospects for agrivoltaics, it becomes evident that this innovative approach has the power to revolutionize the way we approach food production and energy generation.

The transformative impact of agrivoltaics extends beyond improving crop yields; it contributes to reducing greenhouse gas emissions, enhancing water management, and fostering energy self-sufficiency in agricultural communities. As innovative projects continue to emerge worldwide, the momentum behind agrivoltaics grows stronger. By harnessing the untapped potential of agrivoltaics, we can pave the way towards a more sustainable and resilient future for agriculture. Stay tuned as we uncover more exciting developments in the world of agrivoltaics and its invaluable contribution to sustainable agriculture.

Chapter 7: Solar Panel Design and Installation for Agrivoltaics

As the field of sustainable agriculture continues to evolve, harnessing the power of the sun through agrivoltaics has emerged as a promising solution. By combining solar energy generation with agricultural practices, farmers can not only meet their energy needs but also create a more sustainable and resilient farming system. In this chapter, we will delve into the practical aspects of designing and installing solar panels specifically tailored for agrivoltaic setups.

One of the key considerations when designing solar panels for agrivoltaics is the type and orientation of the panels. Photovoltaic (PV) panels come in various forms, including monocrystalline, polycrystalline, and thin-film panels. Each type has its advantages, so selecting the most suitable option depends on factors such as efficiency, cost, and available space. Additionally, the orientation of the panels plays a crucial role in optimizing energy production. Proper alignment with the sun's path ensures maximum exposure, increasing overall output.

While designing an agrivoltaic system, it is pivotal to account for the specific needs and constraints of the agricultural operation. Understanding the layout of crops, the height of vegetation, and the shading patterns throughout the day is crucial for determining the ideal positioning of solar panels. By carefully considering these factors, farmers can maximize energy production without compromising crop growth or yield.

An essential aspect of solar panel installation for agrivoltaics is the choice of mounting system. Mounting structures should be sturdy enough to withstand the elements while providing flexibility for adjustments and maintenance. Common options include ground-mounted systems, pole-mounted systems, and canopy-mounted systems. Each option has its advantages, such as ease of access for maintenance, land utilization efficiency, and adaptability to different crop layouts.

Effective management of the electrical infrastructure is another critical factor in agrivoltaic solar panel installations. Safety measures must be in place, including proper grounding and protection against overcurrent and overvoltage. Additionally, efficient wiring and interconnection systems ensure smooth energy flow and minimize energy losses. Regular inspections and maintenance are vital to identify any potential issues or malfunctions and ensure the continuous and safe operation of the system.

To enhance the efficiency of agrivoltaic systems, innovative technologies such as dual-axis solar tracking can be employed. These systems allow solar panels to automatically adjust their position to face the sun throughout the day, maximizing energy production. While the initial cost may be higher, the increased output can significantly offset the investment in the long run.

In conclusion, designing and installing solar panels for agrivoltaics requires careful planning and consideration of multiple factors. By selecting the appropriate panel type and orientation, understanding the agricultural layout, choosing the right mounting system, and implementing efficient electrical infrastructure, farmers can unlock the full potential of the sun's power. In the second half of this chapter, we will delve into the integration of solar panels with agricultural practices, exploring how agrivoltaics can create a symbiotic relationship between energy production and sustainable farming techniques. Stay tuned for the exciting insights and practical recommendations that lie ahead.A key aspect of agrivoltaic systems is the integration of solar panels with agricultural practices, creating a harmonious relationship between energy production and sustainable farming techniques. By optimizing the use of land, water, and sunlight, these systems can enhance agricultural productivity while significantly reducing greenhouse gas emissions. In the second half of this chapter, we will explore the various ways in which solar panels can be seamlessly integrated into farming operations, offering practical insights and recommendations for maximizing the potential of agrivoltaics.

One of the fundamental considerations when integrating solar panels with agriculture is the careful selection of compatible crops. Certain crops thrive in the partial shade provided by solar panel arrays, making them ideal choices for agrivoltaic systems. Examples include leafy greens, such as lettuce and spinach, as well as herbs like basil and parsley. These crops have lower sunlight requirements and can benefit from the cooling effect provided by the shade of the panels, reducing evaporation rates and conserving water.

Crop spacing and layout are also critical factors to consider for successful integration. By strategically arranging the crops, farmers can optimize sunlight distribution, ensuring that each plant receives the necessary amount of light for photosynthesis. It is advisable to maintain adequate spacing between crop rows and panels to prevent shading and allow for efficient maintenance access. However, farmers should also be mindful of wind patterns, as panels can act as windbreaks, influencing air circulation and potentially affecting crop pollination.

Additionally, agrivoltaic systems offer additional benefits beyond energy production. The shade provided by solar panels can help mitigate the impacts of extreme weather conditions, such as heatwaves, by reducing crop stress and preventing sunburn. Furthermore, the panels act as a protective barrier, shielding crops from hail and other potential environmental hazards. These advantages contribute to increased crop resilience and can lead to higher yields and improved product quality.

Maintenance of both the solar panels and the crops is crucial for the long-term success of agrivoltaic systems. Regular monitoring is necessary to ensure that the panels are functioning optimally, with any dust or debris promptly removed to maximize energy production. It is essential to establish proactive maintenance protocols, including routine cleaning, inspection of electrical connections, and the replacement of any damaged components. Similarly, regular crop management practices, such as irrigation, fertilization, and pest control, should be applied with consideration for the presence of solar panels, ensuring compatibility and avoiding any potential impacts on solar energy production.

As with any agricultural practice, continuous monitoring and data collection are vital for evaluating the performance of agrivoltaic systems. By tracking energy production, water usage, and crop yields, farmers can fine-tune their systems and identify any areas for improvement. Analyzing this data over time allows for informed decision-making, enabling farmers to optimize their agrivoltaic setups for increased sustainability, profitability, and resilience.

In conclusion, the integration of solar panels with agricultural practices through agrivoltaics presents an innovative and sustainable approach to meeting energy needs while maintaining food production. By carefully selecting compatible crops, optimizing crop layout and spacing, and implementing proactive maintenance protocols, farmers can unlock the full potential of this symbiotic relationship between solar energy generation and sustainable agriculture. The second half of this chapter has provided practical insights and recommendations to guide farmers in harnessing the power of the sun for a greener, more resilient future.

Chapter 8: Crop Selection and Management in Agrivoltaics

As the world continues to seek sustainable solutions for agriculture, agrivoltaics, the practice of cultivating crops under solar panels, has emerged as a promising approach. By combining solar energy production with agriculture, we have the opportunity to unlock the true power of the sun and revolutionize sustainable farming practices. In this chapter, we will delve into the intriguing world of crop selection and management in agrivoltaics, shedding light on suitable crops and their specific management requirements.

In the realm of agrivoltaics, careful consideration must be given to the selection of crops that can thrive in the shade provided by solar panels. While some delicate plants may struggle with reduced sunlight, certain species have shown remarkable adaptability in this unique growing environment. One such crop is lettuce, which has been proven to flourish under partial shading. Lettuce plants in agrivoltaic systems exhibit increased leaf size and enhanced flavor due to reduced exposure to direct sun radiation. This, in turn, opens up avenues for optimized lettuce production, where growers can achieve substantial yields while conserving energy.

However, lettuce is not the only crop that holds promise in agrivoltaics. Other leafy greens like kale, spinach, and Swiss chard have also demonstrated the ability to thrive in partially shaded conditions. These crops not only tolerate reduced light but also benefit from the microclimate created beneath the solar panels, leading to improved water retention and reduced transpiration rates. By harnessing synergies between sunlight and shade, agrivoltaics can introduce a new paradigm in sustainable and efficient vegetable production.

Moving beyond leafy greens, studies have explored the feasibility of cultivating fruit-bearing crops in agrivoltaic settings. Research has shown that some fruit trees, such as apple and pear trees, can adapt to the shade cast by solar panels. These trees tend to exhibit altered growth patterns, directing their branches towards the available light. With proper management, such as strategic pruning and training, it is possible to achieve healthy and productive fruit crops, even under the shade of solar panels. By integrating fruit trees into agrivoltaic systems, farmers can diversify their produce and explore new economic opportunities.

In addition to crop selection, effective management practices are crucial for maximizing agricultural productivity in agrivoltaics. Careful attention must be given to irrigation, as the shade cast by solar panels can reduce water evaporation rates. However, this reduced evaporation can also lead to increased soil moisture levels, which may require adjustments to irrigation schedules. Furthermore, the positioning of crops within the agrivoltaic system becomes vital, as different crops have varying shade tolerance levels. Placing shade-sensitive crops closer to the edges of the panel arrays ensures that they receive sufficient sunlight while maximizing land use efficiency.

In this first half of the chapter, we have explored the intriguing realm of crop selection and management in agrivoltaics. From shade-tolerant leafy greens to innovative approaches with fruit trees, agrivoltaics offers a wealth of possibilities for sustainable agriculture. By harnessing the power of the sun in synergy with crop production, we can envision a future where solar energy and food production go hand in hand. But there is still more to uncover and discover in the second half of this chapter – an exploration that will take us deeper into the realm of agrivoltaics and its potential for unlocking even greater agricultural productivity. Stay tuned for the exciting continuation in the next part!In the realm of agrivoltaics, selecting the right crops is only the beginning of the journey towards sustainable agricultural productivity. Effective management practices play a pivotal role in optimizing crop growth and maximizing the potential of agrivoltaic systems. In this second half of the chapter, we will delve deeper into the various management techniques and considerations that can further enhance crop performance in these unique growing environments.

One important aspect of managing crops in agrivoltaics is irrigation. As mentioned earlier, the shade provided by solar panels can reduce water evaporation rates, potentially leading to increased soil moisture levels. While this can be advantageous in terms of water conservation, it also necessitates adjustments to irrigation schedules. Careful monitoring of soil moisture levels becomes crucial to maintain optimal levels for crop growth. Depending on the crop's water requirements, irrigation frequency and duration may need to be modified accordingly. Implementing precision irrigation systems, such as drip irrigation or soil moisture sensors, can help farmers optimize water usage and prevent over- or under-irrigation.

Another key consideration in agrivoltaic management is the positioning of crops within the system. Different crops have varying degrees of shade tolerance. Thus, strategic placement of shade-sensitive crops closer to the edges of the panel arrays allows them to receive sufficient sunlight while maximizing land use efficiency. By thoughtful arrangement and planning, farmers can create microclimates within the agrivoltaic system that cater to the specific needs of each crop, thus ensuring optimal growth and yield.

Beyond irrigation and positioning, agrivoltaic systems offer additional opportunities for enhanced crop management. The panel arrays can act as a protective shield from adverse weather conditions such as extreme heat, wind, or hail. This can create a microclimate that reduces plant stress and improves overall crop health. Additionally, solar panels can serve as a trellis system for climbing crops like beans or cucumbers, providing support for vertical growth and potentially increasing overall yields.

Promoting biodiversity within agrivoltaic systems is another practice gaining momentum. Introducing companion plants, such as beneficial herbs or flowers, can attract pollinators and beneficial insects, creating a more balanced ecosystem that supports crop growth. These companion plants can help control pests, reduce the need for chemical pesticides, and enhance overall resilience against environmental stresses.

As with any farming practice, monitoring and regular assessment are crucial components of effective management in agrivoltaics. Regular crop inspections and soil analysis help identify any potential nutrient deficiencies, diseases, or pest infestations. By employing integrated pest management techniques and utilizing organic fertilizers, farmers can minimize environmental impacts while maintaining crop health. Additionally, harnessing the power of technology through remote monitoring systems can provide real-time data on climate conditions, soil moisture levels, and even plant health indicators, enabling proactive decision-making for crop management.

In this second half of the chapter, we have delved into the intricacies of crop management in agrivoltaics. From precision irrigation and strategic cropping to leveraging microclimates and promoting biodiversity, farmers can unlock the full potential of agrivoltaic systems to achieve sustainable and efficient food production. By combining solar energy production with agriculture, we pave the way for a future where the power of the sun truly enhances our ability to cultivate abundant, nutritious crops while minimizing our environmental footprint.

As we conclude this chapter, we invite you to consider the possibilities that agrivoltaics offer. The synergy between solar energy and crop production holds incredible promise for transforming the way we approach sustainable agriculture. As we move forward, understanding how to optimize crop selection and management in agrivoltaics will be essential in unlocking even greater agricultural productivity and securing a more food-secure and energy-efficient future.

Chapter 9: Maximizing Energy Efficiency in Agrivoltaics

In the pursuit of sustainable agriculture, the integration of solar energy systems and agricultural practices has gained significant attention. Agrivoltaics, also known as solar sharing or dual land use, is a groundbreaking concept that harnesses the power of the sun to promote both crop production and renewable energy generation. This innovative approach has the potential to revolutionize the way we utilize our land, ensuring a more sustainable and resilient future.

One of the key aspects in the success of agrivoltaic systems lies in maximizing energy efficiency. By optimizing the generation and utilization of solar energy, we can unlock the full potential of this symbiotic relationship between agriculture and renewable energy. In this chapter, we explore various methods to enhance energy efficiency within agrivoltaic systems, including energy storage and intelligent control systems.

Energy storage plays a pivotal role in agrivoltaic systems, as it enables the capture and utilization of excess solar energy. With the integration of advanced battery technologies, surplus energy generated during peak sunlight hours can be stored for later use. This stored energy can then be utilized during periods of low solar generation or even during nighttime, ensuring a continuous power supply for both agricultural needs and the electricity grid. By effectively managing energy storage, agrivoltaic systems can enhance overall energy efficiency and contribute to a more stable and reliable energy infrastructure.

Intelligent control systems offer another avenue for maximizing energy efficiency in agrivoltaics. These systems rely on advanced algorithms and real-time data to optimize the performance and output of both solar panels and crop production. By dynamically adjusting the tilt and orientation of solar panels, these systems can maximize solar energy capture throughout the day, further enhancing power generation. Moreover, intelligent control systems can monitor and analyze environmental conditions, such as temperature, humidity, and crop growth, to ensure optimal resource allocation and minimize energy waste. The integration of artificial intelligence and machine learning algorithms enables these systems to continuously adapt and improve their performance, leading to higher energy efficiency and increased agricultural productivity.

As agrivoltaic systems continue to evolve, the potential benefits are immense. These integrated systems not only offer sustainable energy solutions but also provide opportunities for land optimization and increased agricultural productivity. By deploying energy storage and intelligent control systems, agrivoltaic installations can maximize solar energy utilization, reducing reliance on traditional grid energy sources and minimizing overall environmental impact. The increased energy efficiency not only benefits the farmer by reducing operational costs but also contributes to a more efficient and sustainable agricultural landscape.

In the second half of this chapter, we will delve deeper into the practical implementation and case studies of energy storage and intelligent control systems in agrivoltaics. We will explore the potential challenges, benefits, and outcomes associated with these technologies, shedding light on the transformative power they hold. Stay tuned for a closer look at the real-world applications and outcomes of maximizing energy efficiency in agrivoltaics. Exciting advancements await, so let's continue our exploration into the limitless potential of this awe-inspiring union between agriculture and solar energy.As we continue our exploration into the limitless potential of agrivoltaics, we now delve deeper into the practical implementation and case studies of energy storage and intelligent control systems. Through real-world applications and outcomes, we unveil the transformative power these technologies hold in maximizing energy efficiency in agrivoltaics.

Energy storage in agrivoltaic systems not only addresses the issue of intermittency but also allows for the integration of renewable energy into the broader grid infrastructure. With advancements in battery technologies, such as lithium-ion, flow batteries, and thermal storage, the effective capture and utilization of excess solar energy become increasingly feasible. By storing surplus energy during periods of peak sunlight, agrivoltaic systems can ensure a continuous power supply, not only for their agricultural needs but also to support the electricity grid during times of high demand or low solar generation.

One notable example of successful energy storage implementation in agrivoltaics is the System for Integrated Agricultural Solar Power Production (SIASPP). Deployed in Japan, SIASPP combines solar panels, traditional agriculture, and advanced energy storage to create a sustainable and self-sufficient model. The excess electricity generated through solar panels is stored in batteries, providing power for both the agricultural operations and nearby communities during non-solar hours. This integrated approach promotes energy resilience and reduces dependence on conventional energy sources, resulting in both economic and environmental benefits.

Intelligent control systems are another key aspect of maximizing energy efficiency in agrivoltaics. These systems leverage advanced algorithms and real-time data to optimize the performance and output of both solar panels and crop production. By dynamically adjusting the tilt and orientation of solar panels, intelligent control systems can ensure maximum solar energy capture throughout the day, further enhancing power generation. Additionally, these systems monitor and analyze environmental factors like temperature, humidity, and crop growth, enabling optimal resource allocation and minimizing energy waste.

The implementation of artificial intelligence (AI) and machine learning algorithms within intelligent control systems showcases the tremendous potential for improving energy efficiency in agrivoltaics. AI can analyze vast amounts of data and make precise predictions, allowing for dynamic adjustments to maximize both energy production and crop yield. Machine learning algorithms continuously learn from real-time conditions, enabling these systems to adapt and improve their performance over time. By harnessing the power of AI, intelligent control systems can optimize solar energy utilization while minimizing resource consumption, resulting in enhanced energy efficiency and increased agricultural productivity.

A remarkable example of intelligent control system implementation is the "SolGrader" developed by the Fraunhofer Institute for Solar Energy Systems. This AI-powered system uses image recognition and machine learning to analyze crop growth and solar irradiance. By precisely controlling the movements and positioning of solar panels, the SolGrader optimizes both energy generation and crop yields. This innovative technology demonstrates the potential for agrivoltaics to achieve higher levels of efficiency, benefiting both farmers and the environment.

As the field of agrivoltaics continues to grow, the potential benefits become increasingly clear. These integrated systems not only offer sustainable energy solutions but also provide opportunities for land optimization and increased agricultural productivity. By deploying energy storage and intelligent control systems, agrivoltaic installations can maximize solar energy utilization, reduce reliance on traditional grid energy sources, and minimize the overall environmental impact. The increased energy efficiency not only benefits the farmer by reducing operational costs but also contributes to a more efficient and sustainable agricultural landscape.

In conclusion, through the practical implementation of energy storage and intelligent control systems, agrivoltaic systems are paving the way for a more sustainable and resilient future. Real-world examples demonstrate the transformative power these technologies hold, offering energy resilience, economic benefits, and ecological advantages. As we witness the undeniable potential of agrivoltaics in maximizing energy efficiency, we continue to shape a future where agriculture and solar energy intertwine harmoniously, unlocking the true power of the sun.

Chapter 10: Environmental Impacts and Sustainability of Agrivoltaics

Assessing the ecological implications, this chapter examines the environmental impacts and long-term sustainability of agrivoltaics as a renewable energy solution. The harmonious coexistence of solar panels and agricultural practices in an agrivoltaic system holds immense potential for a sustainable future. By understanding the environmental implications, we can better appreciate the benefits and address the challenges that arise from this innovative approach.

One key aspect to consider when evaluating agrivoltaics is land use efficiency. Traditional solar installations occupy large areas of land, often leading to habitat loss and fragmentation. In contrast, agrivoltaics maximize land use efficiency by utilizing the same space for both energy production and crop cultivation. By sharing the land, agrivoltaic systems reduce the need for additional space and prevent the conversion of natural habitats into solar farms.

In addition to maximizing land use, agrivoltaics offer numerous environmental benefits. Solar panels not only generate clean energy but also act as a shield against harsh weather conditions, such as extreme temperatures, frost, and hail. This protection enhances the overall resilience of crops, leading to increased productivity and reduced risk of crop failure. Furthermore, the shading effect of the solar panels reduces water evaporation from the soil, conserving water resources in regions where water scarcity is a concern.

The positioning of solar panels above crops also creates a microclimate that promotes biodiversity and ecological balance. The shading provided by the panels reduces heat stress on plants, creates cooler and moister environments, and slows down evapotranspiration rates. These conditions favor the growth of shade-tolerant plants, promote the presence of beneficial insects, and mitigate the negative impacts of pests and diseases. Thus, agrivoltaics support the preservation of biodiversity and the maintenance of ecological interactions within agricultural systems.

Sustainability lies at the core of agrivoltaics, as it offers a renewable energy solution with minimal ecological footprint. By generating clean energy and reducing greenhouse gas emissions, agrivoltaics contribute to mitigating climate change. Additionally, integrating solar panels into agricultural landscapes increases the overall resilience of food production systems, making them better equipped to adapt to changing climatic conditions. This resilience fosters food security and strengthens the capacity of communities to withstand environmental challenges.

However, as with any farming system, agrivoltaics are not without challenges. The increased complexity of managing both solar energy production and crop cultivation requires careful considerations. Optimum positioning of panels, selection of compatible crops, and efficient water management are crucial factors that affect the success and sustainability of agrivoltaic systems. Additionally, the choice of solar panel materials and disposal methods must align with best practices for environmental stewardship.

In conclusion, agrivoltaics offer a promising pathway to harness the power of the sun while promoting sustainable agriculture. Through maximizing land use efficiency, enhancing productivity, preserving biodiversity, and contributing to climate change mitigation, agrivoltaic systems have the potential to revolutionize the way we produce energy and food. The second half of this chapter will delve deeper into the practical implementation, challenges, and future prospects of agrivoltaics. Stay tuned to uncover the transformative power of this innovative approach.

Agrivoltaics, with their unique combination of solar panels and agricultural practices, hold great promise for a sustainable future. In the first half of this chapter, we explored the environmental benefits and land use efficiency of agrivoltaics. Now, we will delve deeper into the practical implementation of agrivoltaic systems, the challenges they present, and their future prospects.

The success of agrivoltaics relies on careful planning and optimum positioning of solar panels. To ensure maximum energy production and crop yields, experts must consider factors such as solar panel tilt, height, and spacing between panels. Additionally, the compatibility between solar panels and crops is crucial. Some crops can thrive under the shading provided by solar panels, while others may require specific adaptations to grow effectively in this unique microclimate. Ongoing research is focused on determining the best panel configurations and crop combinations to optimize energy and food production within agrivoltaic systems.

Efficient water management is another critical factor in the sustainability of agrivoltaics. The shading effect of solar panels reduces water evaporation from the soil, minimizing water loss. However, it is essential to ensure that crops receive adequate irrigation, especially in regions prone to water scarcity. Employing technologies such as precision irrigation systems can help monitor and deliver water to crops with maximum efficiency.

The materials used in solar panels also play a vital role in the sustainability of agrivoltaic systems. The choice of materials should align with best practices for environmental stewardship, taking into account factors such as resource depletion, waste generation, and pollution. In recent years, there have been significant advancements in solar panel technologies, with increased emphasis on developing more eco-friendly materials and recyclable components. By adopting sustainable solar panel technologies, agrivoltaic systems can further minimize their ecological footprint and contribute to a more circular economy.

Looking towards the future, agrivoltaics have the potential to transform not only energy and food production but also landscapes and communities. As the demand for renewable energy continues to grow, the integration of solar panels into agricultural landscapes offers a unique opportunity to decentralize energy production. This decentralized approach reduces transmission losses and increases the overall resilience of energy systems, making them less vulnerable to disruptions.

Furthermore, agrivoltaics can contribute to rural development by diversifying income streams for farmers. By harnessing renewable energy on their land, farmers can generate additional income through power purchase agreements or by selling excess energy back to the grid. This financial stability can help farmers withstand economic fluctuations and provide new opportunities for sustainable entrepreneurship.

In conclusion, agrivoltaics represent a groundbreaking approach to renewable energy and sustainable agriculture. Continuing advancements in technology and research are paving the way for the widespread implementation of these systems. By maximizing land use efficiency, enhancing productivity, preserving biodiversity, and mitigating climate change, agrivoltaics offer a promising pathway towards a more sustainable and resilient future.

As we move forward, it is crucial to address the challenges and ensure that agrivoltaic systems are implemented with the utmost care and consideration. Through ongoing research, collaboration between different sectors, and supportive policies, we can unlock the full potential of agrivoltaics and create a more sustainable and prosperous world for generations to come.

Stay tuned to uncover the transformative power of this innovative approach and explore the possibilities of agrivoltaics in the practical realm. Together, we can unlock the power of the sun and revolutionize the way we produce energy and food, paving the way for a brighter and more sustainable future.

Chapter 11: Economic and Policy Considerations of Agrivoltaics

As the world increasingly seeks ways to combat climate change and transition to sustainable energy sources, agrivoltaics has emerged as a promising solution that intertwines agriculture and solar energy generation. In this chapter, we delve into the economic viability of agrivoltaic systems and discuss relevant policies and incentives that promote their adoption. By combining solar panels with agricultural practices, agrivoltaics not only harnesses the power of the sun but also offers multiple benefits for farmers and the environment.

First and foremost, agrivoltaics presents a unique opportunity for farmers to diversify their income streams. By installing solar panels on their land, farmers can generate electricity and sell it back to the grid, creating an additional revenue source. This can prove particularly beneficial in regions with high solar irradiance, where solar generation potential is substantial. Moreover, agrivoltaics can mitigate the financial risks associated with traditional farming by providing a stable and consistent income, regardless of seasonal fluctuations or market volatility.

In addition to financial benefits, agrivoltaics offers numerous environmental advantages. By implementing solar panels above crops, farmers can effectively reduce water evaporation rates and provide shade, leading to enhanced water conservation and improved crop yields. This symbiotic relationship between solar panels and agricultural practices creates a microclimate that fosters plant growth, reduces the need for excessive irrigation, and protects crops from harsh weather conditions. Furthermore, the land shared between solar panels and crops is efficiently utilized, maximizing the agricultural potential of limited space.

To foster the widespread adoption of agrivoltaics, governments and policymakers play a vital role in creating favorable conditions and incentives. In many countries, feed-in tariffs and net metering policies have been implemented to encourage renewable energy generation. These policies allow farmers to sell the excess electricity they generate back to the grid at a fair price, thereby incentivizing the deployment of agrivoltaic systems. Furthermore, tax credits, grants, and subsidies can offset the upfront costs associated with installing solar panels, making agrivoltaics financially accessible for farmers.

In terms of agricultural policy, incorporating agrivoltaics into existing agricultural schemes can significantly strengthen the resilience and sustainability of farming practices. By integrating agrivoltaics into national agricultural policies, governments can promote the coexistence of solar energy development and agricultural production. This not only ensures a secure energy supply but also supports local food production, mitigates land use conflicts, and protects valuable agricultural resources. Encouraging research and development in agrivoltaics can also lead to the creation of tailored policies that address the specific challenges and opportunities associated with this innovative approach.

As we have explored in this first half of the chapter, agrivoltaics has the potential to revolutionize sustainable agriculture while contributing to renewable energy generation. Farmers can benefit from diversified income streams and increased crop productivity, while the environment gains from reduced water consumption and enhanced land use efficiency. Through appropriate economic mechanisms and supportive policies, governments can facilitate the wide-scale adoption of agrivoltaics, thereby unlocking the power of the sun for a brighter and more sustainable future.

In the second half of this chapter, we will delve further into the economic and policy considerations of agrivoltaics, exploring additional factors that contribute to its feasibility and success.

One key economic aspect to consider is the potential for job creation in the agrivoltaics industry. As farmers adopt this innovative approach, new job opportunities can arise in various sectors such as solar panel installation, maintenance, and monitoring. Additionally, the integration of agrivoltaics into existing agricultural practices can create demand for skilled workers who understand both solar energy and farming techniques. This not only promotes local employment but also fosters knowledge exchange and skills development in rural communities.

Furthermore, agrivoltaics can have a positive impact on energy security and stability. By diversifying the energy mix, countries can reduce their dependence on fossil fuel imports and enhance their energy self-sufficiency. This is particularly relevant in regions where access to traditional energy sources is limited or costly. Agrivoltaics offers a reliable and renewable energy solution, ensuring a consistent power supply for local communities and industries.

Adopting agrivoltaics also contributes to greenhouse gas emissions reduction and environmental sustainability. By generating clean energy on farmland, carbon emissions associated with traditional energy generation can be significantly reduced. This aligns with global efforts to combat climate change and meet emission reduction targets. Policies that incentivize the adoption of agrivoltaics can contribute to a greener economy and a more sustainable future.

In terms of policy considerations, it is crucial to ensure that relevant regulations and standards are in place to support the development and operation of agrivoltaic systems. Clear guidelines on land use, technical requirements, and grid integration are essential to facilitate the smooth implementation of these systems. Governments and regulatory bodies can work in collaboration with farmers, energy experts, and environmental organizations to establish comprehensive frameworks that address potential challenges and promote the harmonious coexistence of agriculture and solar energy generation.

Additionally, ongoing research and development are essential to continuously improve the performance and efficiency of agrivoltaic systems. Policies that support research institutions, universities, and private companies in conducting studies on agrivoltaics can accelerate technological advancements, paving the way for more cost-effective and productive systems. This research should focus on optimizing panel placement, crop selection, and irrigation techniques to maximize energy generation and agricultural productivity simultaneously.

Moreover, partnerships between energy companies and farming communities can play a pivotal role in the successful implementation of agrivoltaics. Collaboration and knowledge-sharing can enable the tailoring of energy contracts to the specific needs and realities of farmers. These partnerships can further encourage investments in agrivoltaic projects by providing financial security and technical expertise.

In conclusion, agrivoltaics holds immense potential for transforming agriculture and renewable energy generation. The economic benefits include diversifying farmers' income streams, creating jobs, and enhancing energy security. Additionally, the environmental advantages encompass reduced water consumption, increased land use efficiency, and lower carbon emissions. To unlock the full potential of agrivoltaics, governments should implement policies that support the adoption, research, and development of this innovative approach. By fostering collaboration, enacting favorable economic mechanisms, and establishing clear regulations, we can pave the way for widespread adoption of agrivoltaic systems and unlock the power of the sun in achieving a sustainable future for all.

In the ever-changing landscape of sustainable agriculture, advancements and innovations continue to push the boundaries of what is possible. As we look to the future, the field of agrivoltaics holds immense promise. This chapter explores potential advancements and emerging technologies in agrivoltaics that could shape the future of sustainable agriculture.

One area of focus lies in the development of more efficient solar panels specifically designed for agrivoltaic systems. Traditional solar panels may not be optimized for the unique requirements of crop production. Researchers are now working on designing solar panels that not only generate electricity but also allow sufficient sunlight penetration for crops beneath. These innovative panels will have the potential to greatly enhance the productivity and viability of agrivoltaic systems.

Furthermore, advancements in nanotechnology are opening up new possibilities for agrivoltaics. Nanomaterials offer the potential to improve the efficiency of solar panels by capturing a broader spectrum of sunlight and converting it into electricity. Additionally, nanomaterials could be utilized to create transparent coatings that protect crops from extreme weather conditions while allowing the necessary amount of sunlight to filter through. This could help optimize crop growth and protect against potential yield losses.

The integration of new technologies, such as artificial intelligence (AI), into agrivoltaic systems is another exciting prospect. AI algorithms can analyze various factors such as weather patterns, soil moisture, and crop health, allowing for real-time adjustments in solar panel positioning and irrigation schedules. By harnessing the power of AI, farmers can achieve optimal energy production and crop growth, resulting in increased efficiency and sustainability.

The future of agrivoltaics also lies in the development of innovative farming practices that optimize resource utilization. For instance, the utilization of precision agriculture techniques, such as site-specific nutrient management and variable rate irrigation, can maximize crop yields while minimizing resource waste. These practices, when combined with agrivoltaics, have the potential to revolutionize the way we approach sustainable farming.

Additionally, advancements in robotics and automation could have a profound impact on agrivoltaic systems. Robots equipped with sensors and cameras can monitor crop health, detect pests or diseases, and even perform precise tasks such as targeted spraying or harvesting. By reducing the need for human labor, these technologies can enhance efficiency and reduce costs, making sustainable agriculture more accessible and economically viable.

As we delve into the realm of future prospects and innovations in agrivoltaics, it is clear that this field is brimming with potential. The integration of optimized solar panels, nanotechnology, AI algorithms, precision agriculture, and robotics promises to unlock even greater efficiency, productivity, and sustainability in the agriculture sector. The second half of this chapter will delve deeper into these advancements as we explore the fascinating future of agrivoltaics. Stay tuned to uncover the possibilities that lie ahead and their profound impact on sustainable agriculture.

In the first half of this chapter, we explored some of the potential advancements and emerging technologies in agrivoltaics that hold immense promise for the future of sustainable agriculture. From optimized solar panels to nanotechnology, artificial intelligence to precision agriculture, the possibilities are vast. Now, let us delve deeper into these advancements and their profound impact on the agriculture sector.

One area of interest in agrivoltaics is the development of smart grids and energy storage systems. As the integration of solar panels and agriculture becomes more widespread, efficient management and distribution of energy generated from these systems becomes crucial. Smart grids can intelligently monitor and balance energy production and consumption, optimizing the utilization of renewable energy sources. Additionally, energy storage technologies such as batteries and capacitors can help store excess energy during peak production periods and release it during periods of low sunlight or high demand. These advancements not only enhance energy efficiency but also provide a stable and reliable power supply for both the farm and the broader community.

Furthermore, the use of advanced sensors and data analytics has the potential to revolutionize agrivoltaic systems. Sensors can be deployed to constantly monitor environmental parameters such as soil moisture, temperature, and humidity. This real-time data, combined with sophisticated algorithms, can provide valuable insights into crop growth patterns and the health of the system as a whole. By precisely managing irrigation, nutrient application, and shading, farmers can optimize plant health and productivity while minimizing resource waste. This data-driven approach not only improves crop yields but also enables proactive pest management and disease prevention.

Another aspect of the future of agrivoltaics lies in the integration of vertical farming techniques. Vertical farming involves growing crops in vertically stacked layers, utilizing artificial lighting and hydroponic systems. When combined with agrivoltaic systems, vertical farming can maximize land use efficiency while reducing water consumption and eliminating the need for pesticides. This innovative approach allows crops to be grown year-round, independent of weather conditions, and in urban areas where arable land is limited. The integration of vertical farming with agrivoltaics presents the possibility of creating self-sufficient, sustainable ecosystems that provide fresh produce to local communities.

Additionally, ongoing research is focused on developing bio-inspired agrivoltaic systems that mimic natural ecosystems. By incorporating biodiversity into agrivoltaic designs, researchers aim to create not just energy-producing units but also habitats for beneficial plant pollinators, birds, and insects. This ecological approach enhances biodiversity, improves soil health, and promotes natural pest control. These bio-inspired systems have the potential to create a harmonious balance between renewable energy production and ecological sustainability, contributing to resilient and thriving agricultural landscapes.

In conclusion, the future prospects and innovations in agrivoltaics are incredibly exciting. From smart grids and energy storage systems to advanced sensors and data analytics, vertical farming, and bio-inspired designs, the possibilities are endless. The integration of these technologies and practices has the potential to transform sustainable agriculture, increasing efficiency, productivity, and resilience while minimizing resource waste and environmental impact. By harnessing the power of agrivoltaics, we can unlock the true potential of the sun to feed the world sustainably. As we continue on this journey into the fascinating world of agrivoltaics, let us embrace these advancements with open minds and hearts, for a brighter, greener, and more sustainable future awaits us all.

Dear Reader,

I want to extend my heartfelt thanks to you for taking the time to read this book. Your curiosity and willingness to explore the ideas and insights within these pages mean the world to me.

Writing a book is a journey of passion and dedication, and it's readers like you who make that journey truly meaningful. Whether you were seeking knowledge, inspiration, or simply a good story, your decision to pick up this book is a testament to your thirst for understanding and growth.

I hope that the words you've found within these pages have been enlightening, empowering, and, most importantly, enjoyable. Books have a unique ability to transport us to different worlds, challenge our perspectives, and enrich our lives, and it's my sincere hope that this book has done just that for you.

As an author, I am profoundly grateful for the opportunity to share my thoughts and ideas with you. Your support and engagement mean more than words can express. Please know that your time and attention are cherished, and your feedback, if you choose to share it, is invaluable.

Thank you once again for embarking on this literary journey with me. May the knowledge gained from these pages continue to inspire and guide you in your own journey through life.

With deepest gratitude,

Ben Davis

Thankyou

Thank you for your purchase! If you enjoyed this book, please consider dropping me a review. It take 5 seconds and helps a small business like mine.

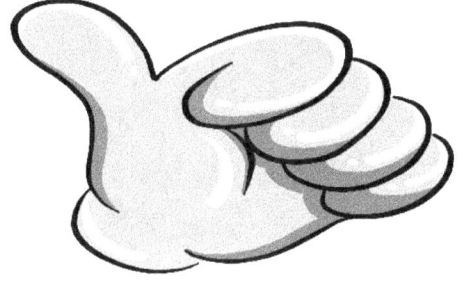